edZOOcation
Big Little Penguin

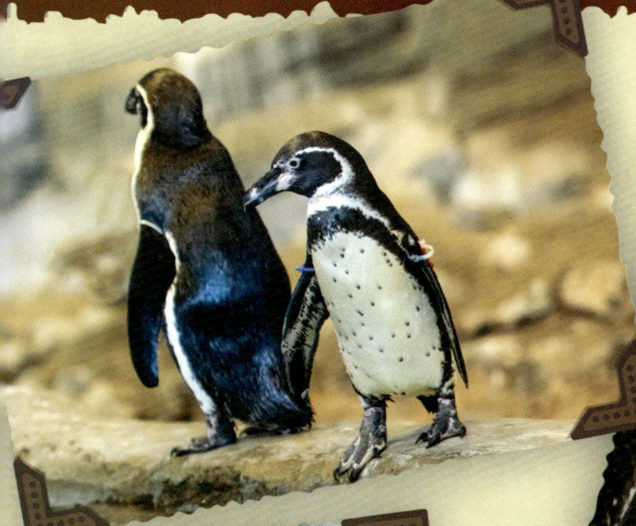

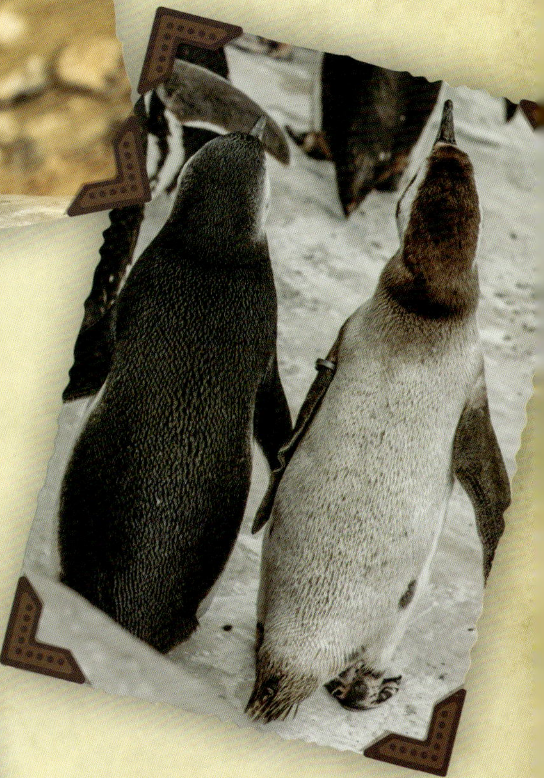

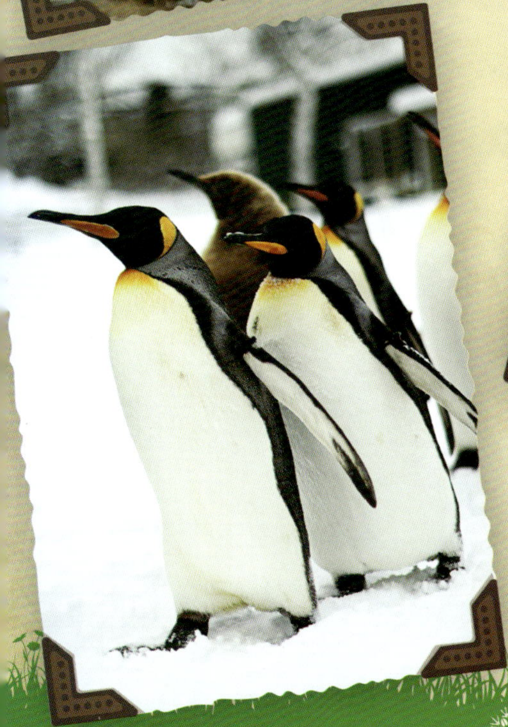

by Sara Karnoscak

Some penguins are very tall.

Some are as tall as you!

Which penguin is the tallest?

Some penguins are very small.

Some are as small as a piece of paper.

Which penguin is the smallest?

Some penguins make nests out of rocks.

It's where they lay their eggs.

Which penguin's nest do you like best?

Sometimes penguins slide on their bellies.

It helps them move faster.

Which penguin do you think is fastest?

Sometimes penguins jump while they swim.

It helps them swim faster.

Which penguin is jumping highest?

Some penguins have crazy eyebrows.

They can be yellow or orange.

Which penguin's eyebrows do you like best?

Some penguins have yellow eyes.

They are called yellow-eyed penguins.

Which penguin's eyes do you like best?

Some penguins have bands around their bellies.

They are called banded penguins.

Which penguin's feathers do you like best?

Some penguins have tails that look like brushes.

Their tails help them swim.

Which penguin's tail do you like best?

Some penguins sleep with one eye open.

It helps them watch for danger.

Which penguins are sleeping?

Goodnight, penguins!

Dedication:

*For parents and caregivers everywhere,
working hard to care for their little chicks.*

—S.K.

For Jordan, who naps like a penguin

—A.R.

Copyright © 2023 Wildlife Tree, LLC. All rights reserved.

Author: Sara Karnoscak

Designer: Allyson Randa

Editor: Tess Riley

Photo Credits:

AdobeStock.com

Pixabay.com

Pexels.com

ISBN: 979-8-9894179-0-2

This book meets **Common Core** and **Next Generation Science Standards.**